NOMORO YA NAANE

THE NUMBER STORY

SMALL BOOK ONE

ENGLISH - SETSWANA

Numbers Teach Children
Their Number Names

written and illustrated by

MISS ANNA

Early Reader Edition of *The Number Story 1*
Bronze Medal Winner, 2016 Wishing Shelf Book Award

Library of Congress Control Number: 2018902040

Names: Miss Anna, author.
Title: Number story : numbers teach children their number names / Miss Anna.
Description: Portland, OR: Lumpy Publishing, 2018.
Identifiers: ISBN 978-1-949320-21-3| LCCN 2018902040
Summary: The pictures and rhymes present stories which introduce numbers 0-10.
Subjects: LCSH Numeration—English—Setswana--Pictorial works--Juvenile literature. | BISAC JUVENILE NONFICTION /
Languages: English—Setswana
Classification: LCC QA141.3 .M57 2018 | DDC 513—dc23

Publisher: Lumpy Publishing
Website: www.missannabooks.com
Email: missanna@missannabooks.com

Paperback: ISBN 978-1-949320-21-3
Printed in the U.S.A.   1 3 5 7 9 10 8 6 4 2

Oka rata go ithutha
ka lebitso la nomoro?

It is very easy and a lot of fun!

E bonolo ene gape e monate!

Say-along our little jingle

Opela le rona bikiyana ka naane ya rona!

starting from Number One!

Retla thomoua ka nomoro ya ntlha!

1

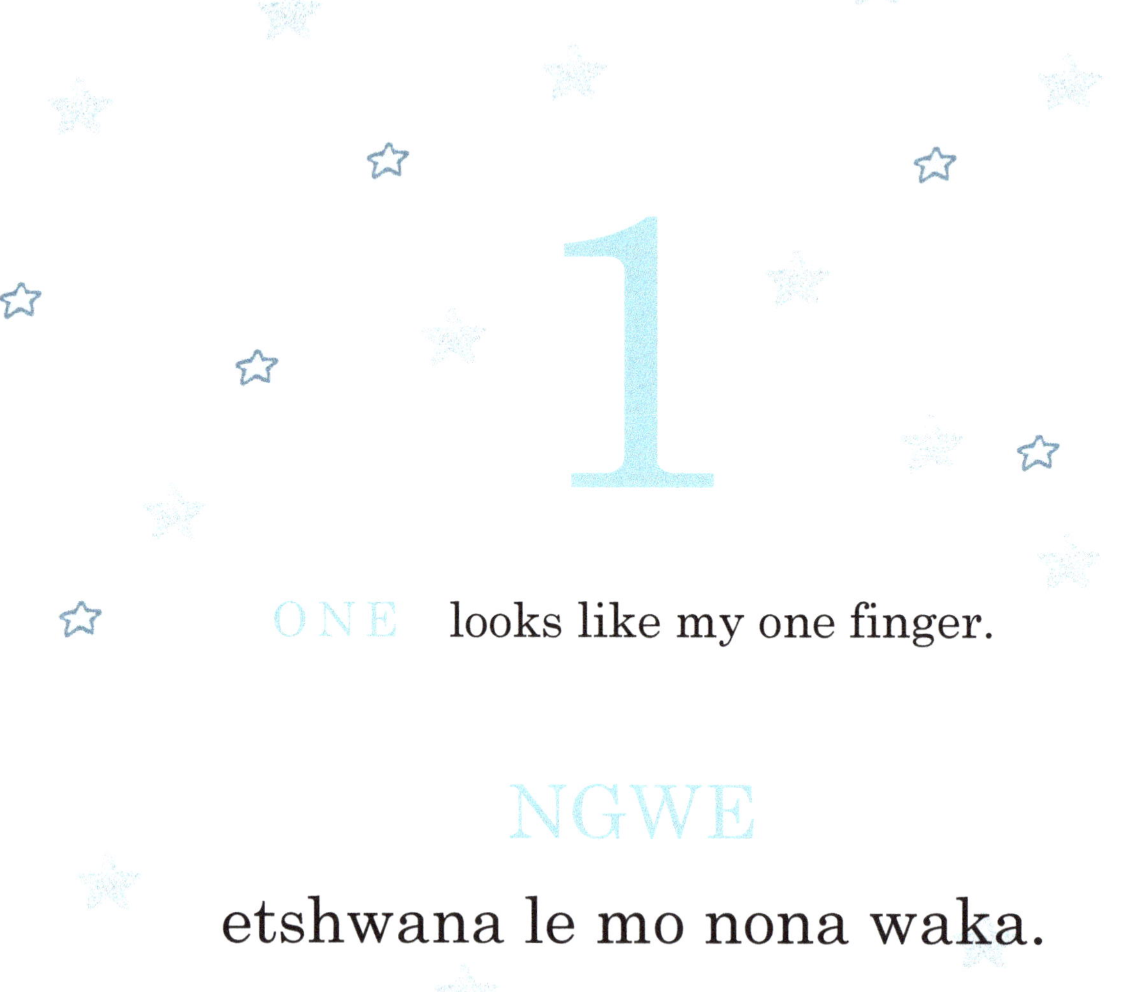

ONE!

ONGEWE!

TWO trails a tail.

PEDI

mogatlha wa letroko.

A TAIL! MOGATLHA!

3

THREE   has bumps.

THARO

e nale bampara.

BUMPY! BAMPARA!

4

FOUR carries a sail.

NNE

e rwala dinete.

4
A SAIL!
DINETE!

5

FIVE   is a racing track.

TLHANO

go thaboga mo tseleng.

VROOM
VRUUUM!
1

6

SIX curves like a snail.

THATARO

khona jaaka feela jaolo sneyl.

A SNAIL! SNEYL!

7
SEVEN has a sharp angle.
SUPA
enale bogale lenyoloi.

BE CAREFUL! IT'S SHARP!
TIHOKOMELA E BOGALE!

8

EIGHT is rollercoaster rails.

ROBEDI

melamu ea rola costa.

YEY!
YIPPEE!

9

 is a bubble on a stick.

ROBONGWE

enale le dipudula mo thobaneng.

A BUBBLE!
DIPUDULA!

10

TEN   is an eye of a whale.

LESOME

leihlo la whale.

WINK!
WINKA!

HELLO! DUMELA!

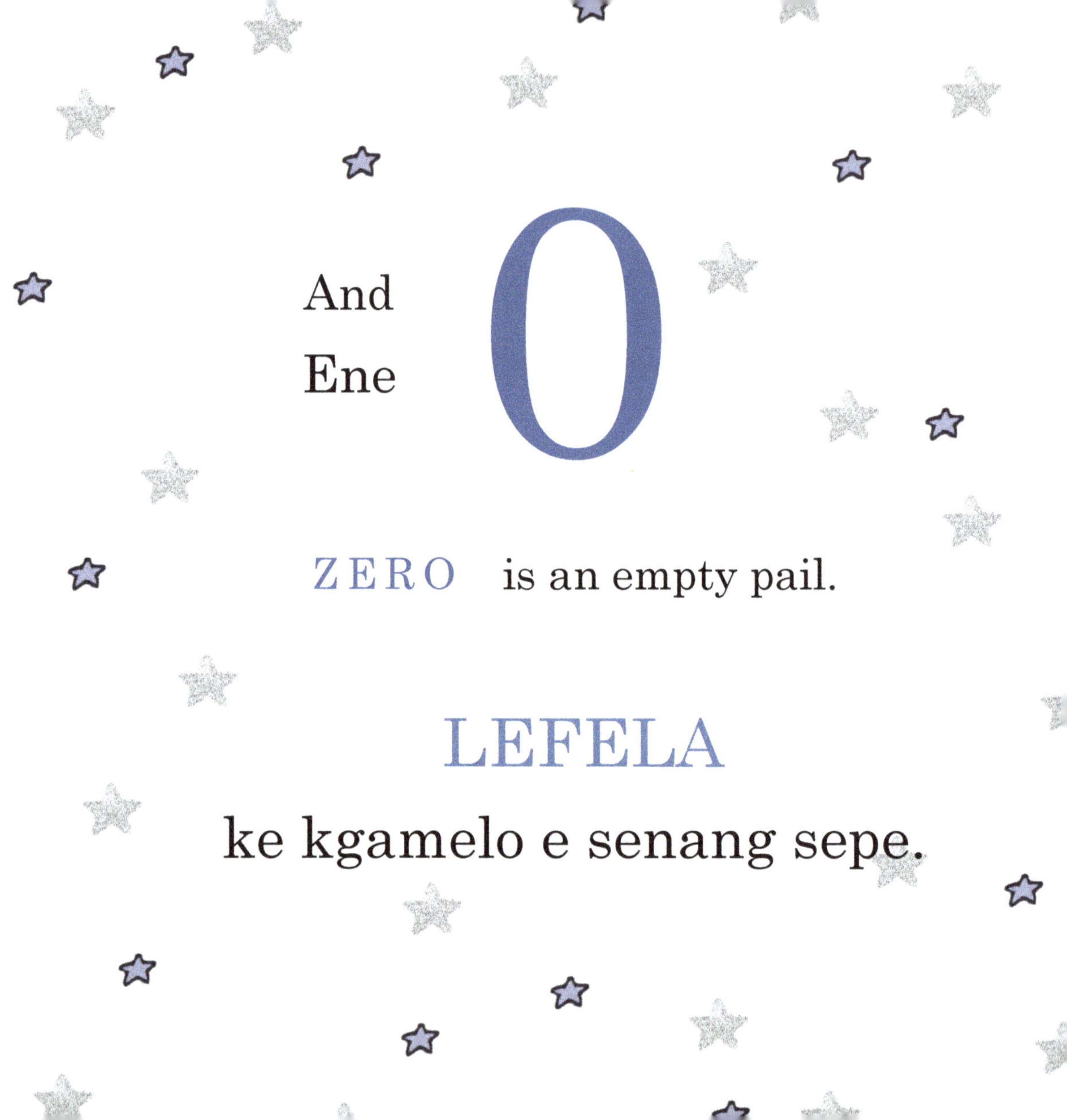

And
Ene

0

ZERO   is an empty pail.

LEFELA

ke kgamelo e senang sepe.

IT'S EMPTY!
GA ENA SEPE!

Thank you for playing with us today.

We had a lot of fun too!

Ke lebogele go tshameka le rona kajeko.

Gonile monate thata thata
fo rona ebe re itumela!

We are your Number friends,
Zero to Ten,
Who will be here for you~
Re dipalo tsa ditsala
Lefela go fitlha kago Lesome.
Retla dula renale wen aka nako tsotlhe.

Bye-bye now!
See you again soon!
Gosiame!
Retlha go bona gare ese kgale!

The Numbers are *SINGING* too!

To sing-a-long, look for Miss Anna Number Story
at your favorite music store like iTUNES.

MP3

Numbers 0-10
IDENTIFYING
& COUNTING

Numbers 11-20
& Ordinals
first, second, third...

Numbers 0-100
& Place Values
ones, tens, hundreds...

About Clocks
& Telling Time
hours, minutes, seconds

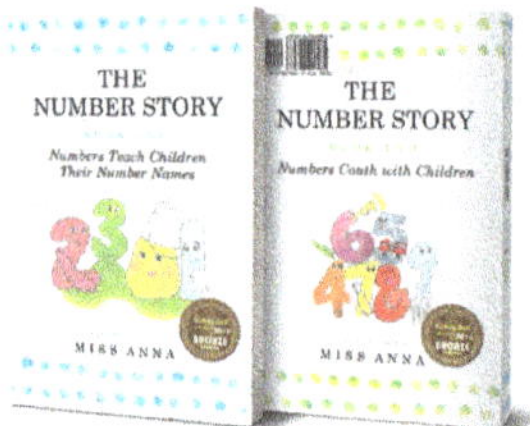

Number Story 1 & 2
isbn: 978-0-996216-48-7

Number Story 3 & 4
isbn: 978-1-945977-01-5

Number Story 5 & 6
isbn: 978-1-945977-06-0

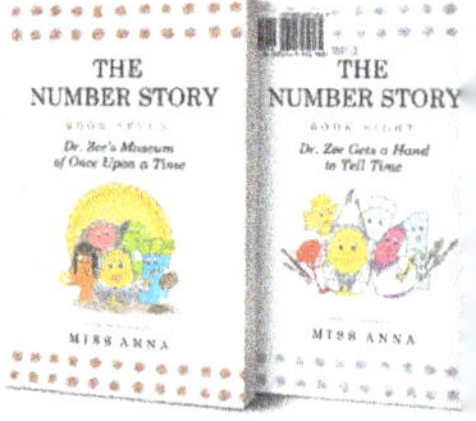

Number Story 7 & 8
isbn: 978-1-949320-40-

For more Miss Anna books to love,
visit us at

www.missannabooks.com

Numbers are working hard all over the world!
Come Travel the World with Us!